Patricia Gordon
304 265-3174

The Author

COVERED BRIDGES
of
WEST VIRGINIA

Past and Present

By Myrtle Auvil

McCLAIN PRINTING COMPANY
PARSONS, WEST VIRGINIA 26287

1972

Standard Book Number: 87012-132-4
Library of Congress Catalog Card Number: 72-85871

Grafton, West Virginia
PRINTED IN THE UNITED STATES OF AMERICA

INTRODUCTION

A bridge is one of man's more symbolic engineering achievements. Below, there are the turbulent waters that threaten to stifle his progress. Across the channel, there is the promise of new and greater things.

West Virginia's covered wooden bridges represent the reaching out of a proud and sturdy people, and the bridges epitomize the strength of their builders.

Huge, oaken timbers, strong like the muscled arms of the mountaineers who built them, stand yet today among the noble West Virginia hills, cradling those who still yearn for something beyond.

It was with proud memories of my early years among these hills that I set out to preserve, in printed form, something of the heritage of these disappearing spans.

Of the fifty-eight wooden structures that once spanned the rivers and streams of West Virginia, in 1972, only eighteen remained. Someday, there may be nothing left to remind us that men once labored long and lovingly to create them.

Most of the covered bridges have served their purpose, and many are no longer used. But because they served so well, they deserve preservation. In an age that often scorns uselessness, they may remind us, by their very permanence, that the values of their builders are also treasures to cherish in a changing society.

If this volume can remind us, in some small way, of a strong and valiant age when men built structures as true and sturdy as their own minds and hearts, then it will have served its basic purpose.

If it does indeed achieve its goal, it will be through the generosity and efforts of many people other than myself.

I am deeply indebted to Edgar J. Harmer of Shinnston, who granted me permission to use material from *Covered Bridges of Harrison County*, written by his uncle, Harvey W. Harmer.

All photos of the bridges now standing were taken by the author.

I am also grateful to the following friends and acquaintances, who donated written material and photographs included in the book:

Dr. Margaret Ballard, of Union; Paul Currence of Napier; the thoughtful librarians in the West Virginia Room at the West Virginia University Library in Morgantown; Mrs. Mary Cornwell, Fairmont; Mrs. Gilbert Reed, Glenville; Charles Wentz of Boulder; Ethel V. Noel, a retired teacher of Monroe, Ohio, who contributed her original poem, "The Covered Bridge"; Virgie Hamilton, the office of the Barbour County State Department of Highways, and Mr. and Mrs. Raymond McDaniel of Grafton, who have furnished transportation for me in my research.

Myrtle Auvil

WHAT OTHERS SAY

"Granddaddy, do you know who taught me how to use a hand planer? It was Miss Auvil who taught me in the elementary school," so my grandson informed me recently.

This exemplifies the life of Miss Myrtle Auvil who served as an elementary teacher in Taylor and Barbour counties for forty years before her retirement in 1961.

Miss Auvil has worked with boys and girls both within the classroom and outside the classroom. For more than forty years she has been active in 4-H Club work. A born naturalist, she has instilled a love for nature in the minds of children who attended her classes.

Although retired Miss Auvil is still very active. She is now serving her third year as president of the Taylor County Retired School Employees organization. She has worked for many years in writing the history and collecting pictures of West Virginia covered bridges. This contribution in depicting the early history of our state will do much in preserving our mountaineer heritage.

Paul C. Bartlett, Superintendent
Taylor County Schools
Grafton, West Virginia

It has always amazed me to watch Miss Auvil as she works, always building toward a brighter future for everyone. Not only in the classroom, where she spent many years helping students over the rough spots, making the work interesting for the children by giving so many things, on the side, that would be invaluable to them for the future.

Miss Auvil was one of the first in Taylor County to see the benefits of the 4-H goals and what their effects would be upon the future life of a person. She has worked incessantly for years to see that these goals have been fulfilled.

In the publishing of this book she has brought back historical and memorable scenes.

Lloyd E. Sayres, Former Principal,
Knottsville School
in Taylor County

I had the privilege of looking over the manuscript of Miss Auvil's book, *Covered Bridges of West Virginia*, just before it was sent to the publisher, and was greatly impressed by her careful research and photography. Her material adds valuable information to the early history of West Virginia highways. Even after the last covered bridge is abandoned, her book will keep alive the memory of these fascinating structures that served the Mountain State so well.

Roy B. Shrout, Past President,
West Virginia Association of
Retired School Employees

THE LURE OF COVERED BRIDGES

Prepared under the auspices of the Mountainaire Travel Council by Margaret B. Ballard, M.D.

The first bridges in America were simply round logs fastened together. There were no side rails or walls. Later trusses were added and then the sides were boxed in. The top of the boxing was covered with flat boards laid on the slant. The direction of the slant could be either toward the floor of the bridge or outward to shed water away from the bridge floor. After a time the bridges were partially covered, particularly the portion that housed the tollhouse keeper's station. Finally the entire structure was covered to protect the timbers from the weather. This last statement is the answer to the often asked question, "Why a covered bridge?"

Many other reasons for building bridges with covers have been given: to provide shelter in the time of rainstorms, to give a young swain a quiet place to steal a kiss from his best girl, to keep horses from shying at the water through the flooring, to keep snow off of the floor, etc.

Covered bridges found other uses than carrying traffic across the streams. They served as shelters for the picnic groups in case of a sudden rain. Circus posters could be nailed on the sides of the bridge and in election years candidates for the various offices found their pictures glaring at them as they traveled around the countryside electioneering. Stories of robbery and murder by highwaymen hiding in the dark recesses grew up around some old bridges. Eric Sloane says in *American Barns and Covered Bridges* that they were used for town gatherings, particularly for meetings the church frowned upon, drill halls for troops and defense in time of war. He also says that many of the old bridges were lit by kerosene lamps.

The best description of a covered bridge in English literature is found in *American Notes*, written by Charles Dickens after his visit to America in 1842.

"We crossed the river by a wooden bridge, roofed and covered on all sides, and nearly a mile in length. It was profoundly dark; perplexed with great beams, crossing and recrossing it at every angle; and through the broad chinks and crevices in the floor, the rapid river gleamed, far down below, like a legion of eyes. We had no lamps; and as the horses stumbled and floundered through this place, toward the distant speck of dying light, it seemed interminable. I really could not at first persuade myself as we rumbled heavily on, filling the bridge with hollow noises, and I held down my head to save it from the rafters above, but that I was in painful dream; for I had often dreamed of toiling through such places, and as often argued, even at the time, 'This cannot be reality.' "

Of all the people who are classed as collectors the folks who collect covered bridges are in a class by themselves. To collect something that you really do not own and cannot carry home is unique. The collectors of rare pianos, antique automobiles, old coins or Staffordshire china may have their collection under their own roofs while the bridge collector must be satisfied with photographs and historical records. It is a recorded fact that over three hundred bridge visitors toured the Mountainaire Travel Council area in 1967. They came from thirty-three states, the Dominion of Canada and three foreign countries. Others wrote for information and photographs of bridges that friends saw in their travels and told them about. The interest grows as the number of covered bridges diminishes.

About fifteen years ago it was estimated that there were nearly two thousand covered bridges in the United States. West Virginia had more than fifty covered bridges in 1954 when Richard Sanders Allen of Round Lake, New York, took a census of the covered bridges in the United States. In a recent folder prepared by the Travel Development Division of West Virginia Department of Commerce twenty covered bridges are listed in West Virginia.

There are four covered bridges in the council area and one just over the line in Pocahontas County. Two of these bridges are in Greenbrier County. One is over Milligan Creek northwest of Lewisburg on Secondary Route 40. Directions: Take U.S. 60 West out of Lewisburg for about three miles, turn left on routes 60/11 and almost directly left again on Secondary 40. It is about one mile to the bridge from this turn. The bridge is in use.

The second bridge in Greenbrier County is over Second Creek near Hoke's Mill. Directions: Approach from the south, take U.S. 219 out of Union, three and one-half miles to Pickaway, turn left onto Secondary 5, keeping to the right-hand road at the forks at the edge of town, routes 5/3. Keep on this road for about two and one-half miles to the bridge. Note the long dry millrace track on the opposite side of Second Creek. The ruins of Hoke's Mill may be seen farther down the creek at the ford. Only the stone foundation remains. Or if you are traveling from the north on U.S. 219 turn right at Organ Cave, three miles south of Ronceverte onto Secondary Route 62. Continue on this road to the bridge, about five miles. You will pass the ruins of Hoke's Mill on the right near the ford of the creek at the bottom of the hill. This bridge is in use.

Indian Creek Restored Covered Bridge is in sight of U.S. 219, six miles south of Union at Saint John's Methodist Church. There are parking facilitics and a walkway to the bridge. Early vehicles and tools used in bridge building are on display. This bridge is not in use, but is maintained by the Monroe County Historical Society. Watch for markers on the highway.

Laurel Creek Bridge, near Lillydale is in daily use. Directions: Take U.S. 219 out of Union, south of Salt Sulphur Springs, then turn right onto Secondary routes 219/7 to Lillydale. Keep right around church and old store building and continue about five miles to the bridge. This road leads to Greenville and passes en route the large opening to the Laurel Creek Cave near that village. The cave opening is easily seen from the highway on the right.

Locust Creek Bridge in Pocahontas County is near Hillsboro. Directions: When traveling north on U.S. 219 ask directions at Renick or if traveling south inquire at Mill Point.

Even though these bridges, with one exception, are on secondary roads you should reach them without difficulty. We hope that you enjoy your bridge collection experience in the Mountainaire Travel Council area.

THE COVERED BRIDGE

Within the confines of a wood
Where alders and the white birch stood
The covered bridge serves as the link
That spans the river, brink to brink.

American as apple pie
It bridges water, roofs the sky.
Symbol of horse and buggy days
Reminder of more primitive ways
Shelter it is from rain or hail,
From winter snows or autumn gale.

How often have we crossed the span
Which seems a part of nature's plan
Clip, clop, clip, clopping went the beat
Of iron shoes on Dobbin's feet.

May no storm rend, may no hand mar
Its weathered timbers, as they are.
To grace the view long may it stand
Loved object in a pleasant land.
Seen first from yonder neighboring ridge
Artist's delight, the covered bridge.

Ethel V. Noel
Retired teacher, Monroe, Ohio

to defend them, right or wrong. One who believed himself right, Bailey Brown, was slain at the very portals of the historic old structure on Tuesday, May 22, 1861.

During the week of June 10, 1888, a week of rain caused an abnormal amount of water to pour down the slopes of the Tygart Valley watershed, causing the greatest flood in the memory of the oldest inhabitants. Floods frequently occurred in the early spring from melting snow that covered the slopes, but for this to happen in midsummer was most unusual. Floodwaters reached the highest mark ever known. Most homes on Maple Avenue were inundated, the one-story buildings covered completely. It was feared the raging waters would sweep the Baltimore and Ohio Railroad bridge from its piers. Heavily loaded freight cars were hastily pushed out over its length to hold it in place.

At Fetterman, the raging waters lifted the old historic covered bridge from its piers and flung it into the swiftly running waters to the stream below and carried it from the place it had stood for more than half a century to forever.

The covered bridge over Sandy Creek at Claude, Taylor County, was a connecting link between Taylor and Barbour counties. It was built in 1893 and 1894. I was informed that the deed to the location was given by John Miller. John E. "Bud" Cozad was a carpenter and helped with the construction. It has been replaced with a concrete structure.—Photo courtesy Robert Goodwin, Grafton.

The pier of the old Fetterman Covered Bridge still stands in the Tygart Valley River.

In 1887 a covered bridge was constructed across Thirteen Mile Creek, at Buxton Mill in Mason County, just above the old mill. It was approximately 150 feet long and stood high above the waters of "Thirteen." An iron bridge replaced the old covered bridge in 1949. The old mill in the picture was built on Thirteen Mile Creek in 1877 by Darius Buxton. The large four-story mill was the center of trade in that area. Picture and history furnished by Mrs. Juanita Burdette of Huntington, West Virginia.

The W. T. Law Bridge across Simpson Creek, near Bridgeport, was built in 1881. Colonel T. S. Spates was appointed by the Harrison County Court in 1881 to locate a site and advertise for building a covered bridge at Holland's Mill. September 9, 1881, he reported that he had secured a site for the bridge on the farm owned by John Lowe. He allowed Lowe $40 for the site and awarded the contract for construction to A. S. Hugill for the price of $974. The court approved the report and on December 6, 1881, Spates reported the bridge had been completed. It is seventy-eight feet long and sixteen feet wide. It withstood the great flood of July 1888, and is still standing in good condition. It is the largest of the three bridges still standing in Harrison County.

Rooting Creek Bridge in Elk District has been used continuously since erected in 1887. Rooting Creek, in Harrison County, is the longest branch of Gnatty Creek. On May 2, 1887, the county court entered an order appointing James M. Eibb a commissioner to examine a bridge site across Rooting Creek near his residence and report to the court the kind of bridge necessary to be built, and the cost of erection. On July 11, 1887, Eibb made his report. The report was approved by the court and they appointed George C. Blair to build the bridge according to the plans, for the sum of $675. On October 6, 1887, Mr. Eibb, who also had been appointed to superintend the construction, reported it had been built by Blair. Blair was paid $675. It has been in continuous use but is not in good condition at present.

The Fletcher Bridge.

THE FLETCHER BRIDGE

The Fletcher Bridge is across Tenmile Creek on the road leading north from Route 50 to Marshville. It derived its name from the well-known family living nearby.

On July 1, 1891, the Harrison County Court appointed John T. Williams a commissioner to advertise and receive bids for a covered bridge across Tenmile Creek, on a site selected by commissioners previously appointed. The bridge was to be built according to plans and specifications on file in the clerk's office. He was required to advertise in a newspaper for fifteen days before July 25, 1891, and make his report not later than the first day of August.

The court appointed William J. Williams as construction superintendent. The contract for the masonry, or abutments, was awarded to L. E. Sturm. On November 27, 1891, William J. Williams reported that the abutments had been completed at the cost of $937.46 and the superstructure was nearing completion. The large stones for the abutments were quarried at the top of the hill about one hundred feet above the west end of the bridge. Some of the stones are about two feet thick and six or seven feet in length.

The bridge is still standing and in good condition. It is sixty-two feet in length; three-ton load limit.

The Kinchelow Bridge, Harrison County, built about 1897, has been replaced by a concrete bridge.

The Locust Creek Bridge was built in 1870. It is on the original road linking Pocahontas and Greenbrier counties. (This was the first road established by the pioneers.) It was repaired in 1904-05. Still in good condition, it is 116 feet long and may be reached on Route 31 at junction of Primary 20 West.

Two Wadestown covered bridges, Monongalia County. These relics of the past gave way to new concrete structures. Mrs. Blanche Shriver and a Mr. Price loaned the negatives.

CHEAT BRIDGE

When people get up in years, they are sometimes given employment in some easy job such as night watchman, or perhaps crossing sweeper, and such seems to have been the case of the old wooden covered bridge that spans Cheat River in Preston County. Located five miles south of Rowlesburg and known as Cheat Bridge, it was retired from duty to vehicular traffic in 1934 upon the completion of a fine modern three-span steel and concrete bridge just a short distance to the north or downstream.

But strange as it may seem, after the old wooden structure had been condemned as unsafe for heavy loads, nearly all the heavy steel and huge construction machinery were hauled through the old bridge to be used in the construction of the new bridge, as the laying of the new bridge was started from the east side of the river from Rowlesburg where the construction supplies were shipped, a beating from which the ninety-nine-year-old structure came up smiling, even though a bit shaky at times when a huge trailer truck carried the massive sills and other structural steel through its eastbound lane, to be used in the new bridge.

In its day, Cheat Bridge, as the 339-foot span has been called, was one of the best pieces of bridge engineering construction in the country. This two-lane bridge was constructed of white pine timbers which were cut at various points in Tucker County, some from the Maxwell place near Saint George, and the large timbers in the arches, of which there are eight, were cut from the old Blackman place just east of Parsons.

These timbers had been cut and skidded to the river, then made into crude rafts, and when the water stage was right for rafting, they were floated downstream to the bridge site. While moored there awaiting the construction crews, the river raised suddenly from hard rains at headwaters, and washed the timbers away and other timbers had to be cut and rafted down as before.

The huge timbers that form the arches that support the weight of the bridge between the piers are mortised together

Its memory lingers on . . .—Photo courtesy Harry Slaughter, Aurora, retired teacher.

without benefit of nail or iron bolts as there were no nails at that time. It was only in later years when repairs were made to the bridge that nails were used in the structure.

According to Wiley's *History of Preston County*, Josiah Kidwell was the contractor. Money was raised for road and bridge construction by a form of lottery or selling shares and the revenue came from tolls after the work was completed to pay off the shares. For many years Cheat Bridge was a toll bridge. Among the persons that kept toll were James J. Hollis, grandfather of H. R. Hollis of Rowlesburg. Other toll-keepers were Dr. Call and Kate Bishop, a daughter of Christopher Bishop. It is unknown just when toll was discontinued, sometime long after the close of the Civil War.

On April 27, 1863, Jones's army of more than two thousand men passed over the bridge and en route they tore up the flooring on the one side of the span and nearly two years elapsed, during which time one way traffic was maintained on the bridge, before the flooring was replaced.

The bridge was always painted red until 1931, when the State Road Commission had it painted battleship gray.

Many of the huge timbers in the old bridge, which is 124 years old, are in a splendid state of preservation and many bear the carved names or initials and dates of passersby. Romance has had its share in the carving, and if the old bridge could talk, its stories would be legion.

Today the sightseers, as of a century ago, continue to leave their letters by carving or painting, as a reminder they have passed this way.

At present this ancient bridge, though retired from heavy employment, basks in its continuing glory, serving as a relic of West Virginia's pioneer past, and invites the public with its rest stations, table for picnics, and the observation view of beautiful Cheat River for more than a mile to the south where boating, bathing and fishing attract many in the summer season, and holds the interest of the traveling public the year round.

It is in need of some repairs, however, especially at the west end, where some of the heavy timbers are showing signs of decay, and one of the piers has been damaged by floodwaters.

A new roof was put on the old bridge a few years ago, and it was hoped the state would look after these other needs at an early date. However, in the summer of 1964, tragedy struck this memorable old edifice, when it was destroyed by fire and fell into the Cheat in twenty minutes.

PHILIPPI COVERED BRIDGE

The Philippi Covered Bridge was built in 1852, the woodwork by Lemuel Chenoweth of Beverly and the stonework by Emmet O'Brien. Mr. O'Brien was then a citizen of Barbour County but later moved to Lewis County. Mr. Chenoweth was a bridge architect by profession, and his designs were original. Every principle was worked out with mathematical precision. He knew beforehand the shape and size of every piece of timber used in the framework of his bridges. Born in 1811 at Beverly, he took an active part in politics and was superior to ordinary men in both education and natural endowments.

In 1850, the Board of Public Works of Virginia was accepting bids for the construction of the major bridges for the Staunton to Parkersburg turnpike. Bids for the contracts had been extensively advertised. Bidders were present in large numbers from the East and North, with all sorts of models and plans, including iron structures, wire cables, cantilevers, stone arches, and wooden models of every kind. Mr. Chenoweth, who had spent many weary hours whittling and cutting a model from hickory wood, as strong as it could possibly be made, was present.

On the appointed day the bidders assembled before the board of public works. So far as appearances went, some of the New England Yankees had models of perfect form and beauty. Mr. Chenoweth, a tall gaunt man with long black hair, was the last called. Tired from his 250-mile trip by horseback, he quickly stepped forward, assembled his model he had carried over the mountains. Placing two chairs face to face, he set his bridge on them, climbed upon it and walked its length. "Gentlemen, this is all I have to say," were the only words he spoke.

Needless to say, this astonishing feat, plus the low bid, saw the contract awarded to Lemuel Chenoweth, who went on to become West Virginia's most distinguished bridge builder. He built others at Beverly, Buckhannon, Weston, Middle Fork and Barrackville.

Today two of his covered bridges are still standing and in

PHILIPPI, W.VA.
SCENE OF FIRST LAND BATTLE OF
CIVIL WAR
This Bridge Erected 1852 Served Both North and South
in Passage of Troops and Supplies Across Mountains into Virginia

Historic covered bridge at Philippi, West Virginia.

regular service, the historical Philippi bridge and one across Buffalo Creek at Barrackville, Marion County, built by him and his brother, Eli Chenoweth.

The Philippi bridge is one of America's most famous bridges. It was here on June 3, 1861, that a small garrison of Southern soldiers, some of them billeted in the bridge itself, were surprised by Northern cannon fire.

The Confederates were driven from the town in the skirmish, recognized as the first land battle of the Civil War.

During the construction of the bridge, typhoid fever broke out among the workmen. There were fifteen to twenty men ill at the same time, which nearly stopped the work, as it was almost impossible to get new workers.

In 1852 the Philippi bridge was opened and a tollgate was placed at the east entrance, between the bridge and the railroad crossing. Some of the prices to cross the bridge were: horse and rider, 10¢; for every led or driven horse, 5¢; for every stagecoach drawn by two or three horses, 45¢; for every stagecoach drawn by four horses, 50¢; for every head of cattle, 1½¢; a score of hogs, 10¢.

During the Civil War, it not only served as barracks for the Union Army, but also wartime traffic flowed over it from the railroad at Webster, as supplies moved south along the Tygart Valley River, to maintain an army of occupation throughout the reaches of the river.

Although many of the covered bridges were destroyed during the Civil War, the Philippi bridge was left intact, largely because it was controlled by the Union forces.

One story is told that the bridge was mined during the entire war so it could be destroyed at any time that it was threatened by the Confederates. Today, after 120 years, this bridge is still an integral part of an important highway.

Spanning the North Fork of the Hughes River near Cairo, in Ritchie County, this bridge was known as the McKinney Bridge. It was built in 1878. The stonework to support this bridge was done by James P. Amick, a stonemason from Virginia. He and Edgar Trainer, a carpenter, did the woodwork on the bridge. This was the first covered bridge built across Hughes River. It was left intact but replaced with a new bridge. The past few years it began to show the marks of neglect and age. Finally, on September 3, 1970, the old landmark collapsed and fell into the river. Information courtesy Ward Gamble, a retired schoolteacher of Harrisville, West Virginia.

Covered bridge at Cairo, after fall into the Hughes River.

The covered bridge over Fish Creek, Hundred, Wetzel County, was constructed about 1892. Relatives of C. W. Critchfield near Hundred helped with the building. It is still in very good condition. It is thirty-six feet in length. It can be seen from U.S. Route 250 South, near Hundred, at the junction of Route 13. (See next page.)

Mrs. C. W. Critchfield shown holding the broadaxe used to cut the logs for the Fish Creek bridge.

INDIAN CREEK COVERED BRIDGE

In my search for information about the Indian Creek bridge, I borrowed a copy of the *Monroe Watchman* for November 5, 1964. It contained a copy of a letter from Oscar Weikel, one of the builders of the bridge.

I wrote to him in Santa Rosa, California, about the use of the article. His reply follows: "Read your letter about the covered bridge. Yes, you have my permission to use my story. I am glad you are taking interest in the bridge and would like to have one of your books and thank you." That was in 1969.

The letter to the *Monroe Watchman:*

"I am more than interested in the old bridge as I was one of the contractors that built the bridge, and was one of the youngest contractors in Monroe County at that time.

"As one of our first contract jobs, my brother, Ray Weikel, aged 18 years and I built the covered bridge. I was 16 years old at the time.

"About the shingles, as Mr. Thompson said they were split and shaved chestnut. The bridge was made of pine lumber. We sawed the timber on the Hill above the old Harvey Baker place on what is now known as the Smithson place. We manufactured all the timber on our own mill; we were the lowest bidder on the bridge and we were teenagers. Our uncles, E. P. Smith Brothers, told the county court they would stand behind us and we were granted the contract, so that made us the youngest contractors in the county, and our first job. I am 81 years of age and the bridge is still standing, close to my heart. To all of the good people of Monroe County and State, get together and save the old landmark.

"I think the covered bridge was built in about 1898 or 1899, and little did we think it would still be standing some sixty years later. Nothing would please me more than to see it preserved. I think I am the only one still living that had anything to do with construction of the bridge. I would be glad to answer any questions about it."

The old bridge still stands, in the field across the highway from the home of H. B. Kessinger. When the new road and

Indian Creek Covered Bridge.

new bridge were constructed, the Monroe Historical Society took on the chore of restoring the bridge. The society furnished the money, along with other contributors. One of those helping with the work was Walter Weikel, age seventy-three, who was still able to swing maul and mallet. He also knew how to split out shingles with the froe, an almost lost art in this day of machine-made material.

Inside the bridge may be seen the froe, maul, mallets, and tools used in the original construction. An old buggy and sleigh may also be seen.

It is well worthwhile for anyone interested in this type of Americana to make a trip to Union, Monroe County, to see this preserved specimen. In the fall, I assure you, you will be thrilled at the beautiful setting.

Maulsby Bridge across the West Fork River near Shinnston knew both romance and tragedy.

MAULSBY BRIDGE

The Maulsby Bridge was the longest and probably best known covered bridge in Harrison County. The Weston and Fairmont Turnpike Company was chartered by the legislature of Virginia by an act passed in 1848 for the purpose of building a road from Weston to Fairmont.

To build the pike from Clarksburg northward to where it would cross the West Fork River, a bridge was built about three miles south of Shinnston. The bridge was built by Elnore Hart of Clarksburg. It was about one hundred yards long and was supported by two abutments and a center pier built of large stones.

The chief object of the company was to make money for the stockholders as well as serve the public, therefore, those using the pike were required to pay toll. In order to collect the toll, gates were installed across the pike at certain points. One of these tollgates was at the east end of this bridge.

The first keeper of the tollgate at the bridge was a man named Maulsby, so the bridge was named for him.

Granville Davisson Hall, in his book, *Daughter of the Elm*, describes a well-planned holdup that took place on this bridge. However, the thieves received no loot since their victim had left his money in Clarksburg. On April 30, 1863, William E. Jones, a general in the Confederate Army, made his celebrated raid up the Monongahela Valley capturing horses and cattle and merchandise.

After passing through Shinnston he divided his forces and sent the larger body up Simpson Creek to Bridgeport, and the smaller body to Clarksburg. The Union forces in Clarksburg, learning of Jones's approach, sent Major Lot Bowen, at the head of some sixty cavalrymen, to head him off and keep him out of Clarksburg. Bowen met the Confederates south of this bridge. During the scrimmage one man was killed on each side. James W. Custer was the Union man and Peter Armstrong was the Confederate.

This bridge over Sandy Creek in Jackson County is 102.5 feet long and was still in fair condition when photographed in 1969; on routes 21/15 at the junction of U.S. 21 East, between Sandyville and Odaville. It was built in 1892. See entrance view of bridge on opposite page.

Entrance view of bridge over Sandy Creek.

The covered bridge at West Union, in Doddridge County, was completed in 1843. This bridge was completed before the Philippi bridge in 1852. A contract was let to a man by the name of Lynch of Harrison County. The timber was whipsawed by old-time axeman Ephraim Bee. A blacksmith made all the bolts and bands for the bridge. The day of opening people came from miles around. A young lawyer of Clarksburg was the speaker. He started as follows: "My fellow citizens, the material for this bridge a few months ago was forests." He repeated the same words a second and a third time, but failed somehow to get them connected with his oration. He bowed his head saying, "I would not care a damn if it was still the mighty forest, folks, after this embarrassing moment." He made a fine talk. In later years he became an excellent lawyer in the Clarksburg area. Middle Island Creek is ninety-two miles long and said to be the longest stream in the United States to be called a creek. The bridge was washed away with the flood of 1950.—Picture and information furnished by Mrs. Paulene B. Roush, West Union, West Virginia.

The little Center Point Covered Bridge, Doddridge County, was abandoned a few years ago when a new road was constructed. Located on Talkington Fork of McElroy Creek, it served the community for many years. Mr. and Mrs. Willie Strader, Sr., told me they made many trips through the bridge by buggy and wagon or sleigh, then later by automobile. It is privately owned now. William Morgan exchanged a right-of-way for the road for the use of the bridge. When photographed it was being used as a garage.

Bridge over Indian Creek at Springfield, Monroe County. No longer standing.

MILTON BRIDGE

Restoration to one of Milton's most historic landmarks, a Cabell County covered bridge, located on the East Mud River Road, was undertaken and completed in 1971 by the State Department of Highways, through the efforts of the Covered Bridge Garden Club of Milton, an organization in existence only three years.

The club was named after the historic bridge, and its twenty-four original members dedicated themselves to having the bridge restored, maintained, and the area eventually landscaped.

The first step taken by the club was on October 21, 1969, when the town council was approached with a request that it have installed two dusk-to-dawn lights, one at each end of the bridge. The club hoped the lighting would cut down on vandalism to the structure. Council voted to install the lights, but when it was learned one light would be out of city limits, it became necessary to appeal to the county court. Court members agreed to help.

The next step was to approach the Department of Highways to ascertain what could be done toward restoring the bridge. Various permits finally were obtained and on July 28, 1970, the club was granted a permit to erect a historical marker near the bridge, erect directional signs, fill in dirt and rocks around the riverbank near the bridge, and install the dusk-to-dawn lights.

Plans for a complete structural restoration of the bridge during 1971 then was instituted and work began in May. Flowers and shrubs will be planted by the club when restoration is completed.

In an effort to establish the actual age of the bridge, club members talked to many of the senior citizens in Milton and Tri-State area. Stories of the bridge have been colorful and various, but no exact date of construction can be pinpointed. Some say it was built prior to the Civil War; others say it came later. According to records of the Department of Highways, it was built in 1821 and first repairs made in 1939-40.

While the actual genealogy of the bridge is uncertain,

members of the Covered Bridge Garden Club hope to continue to trace its history. Old maps, letters, documents from which a history might be extracted are needed. Since there were, at one time, two covered bridges in Milton, there has been a great deal of confusion.

In an old deed book in the county commissioner's office is the following information:

"July 14, 1891—In the matter of the bridge across Mud River just below Milton. It is ordered by the Court that T. A. Love and T. B. Summers be appointed a committee to have the said bridge repaired and in safe condition for travel as soon as possible and to report to the next term of this court and furnish an itemized account of expenses approved by said committee."

"November 6, 1891—In the matter of the Milton Mud River Bridge—T. A. Love and T. B. Summers committee herefore appointed to have said bridge repaired this day made their report to the court, which report being seen and inspected by the court is approved. It is ordered that the several parties hereafter received be allowed the sums opposite their names for services rendered in repairing said bridge to wit:

"Swann, lumber, $11.73; Gerlach, shingles, $38.50; T. A. Love, lumber, etc., $20.00; S. Roberts, sundry articles, $3.86; Yates & Murphy, paint, $5.00; Jacob Williams, labor, $13.00; Jas. Chapman, use, all labor, $5.50; T. B. Summers, labor, $11.00; J. W. Reynolds, black smith, $1.50. It is further ordered that the clerk of this court issue orders to the above parties for said sums payable out of the levy of 1891."

The preceding article, from the *Cabell Record*, Milton, June 23, 1971, as well as the photograph of the bridge being repaired, were given to me on April 12, 1972, by the editor and owner, Caldwell Dudley, on a visit to his office while on a trip to photograph the bridge since its restoration.

A sign on the end of the bridge reads:

BRIDGE MAINTAINED BY
THE UNION ARMY
AGAINST
THE CONFEDERATES

Milton Bridge before repairs.

Covered Bridge, Milton, Cabell County.

The following is through the courtesy of Caldwell Dudley, editor and owner of the *Cabell Record*.

The first semblance of a community settlement in or around the present site of Milton occurred in relation to the establishment of the Kanawha and James River Turnpike, as it passed through this area. This East-West thoroughfare was built in the early 1830s and over it a stagecoach line operated from 1835 to 1873.

Prior to 1834, a ferry across the Mud River was operated by a John Morris of Teays at a site approximately two miles southeast of the present town of Milton.

In 1834 a covered bridge was built at this ferry crossing by Charles Conner of Putnam County. It was made from timber that grew nearby. The material was hewed with the broadax and adz, hauled to the site by oxteams, and put in place by strong and sturdy hands. The bridge was first covered with boards made from white oak and split to shape with a froe. Wooden nails and pegs were used to fasten the boards and the timbers.

Soon after the bridge was completed, a stagecoach stop and inn was opened in a log structure on the west side of the river by John Morris, who later sold this business to Abia Rece. A short time later, an official post office, Mud Bridge, was opened across the road from the inn by Abia Rece's son, Milton Rece, who in turn sold this business to James Everett.

It is believed that the Marquis de Lafayette, Henry Clay, Andrew Jackson and John Breckinridge made use of this route, as did Abraham Lincoln in 1847 when on his way to Washington to take his seat in Congress.

Some of the early settlers of the community at that time were: John Morris, Edmund Morris, John Jordon, Andrew Jordon, Charles Venable, James Beckett, Joseph and Abia Rece, John Everett, Nathan Everett, and Thomas Harmon.

The little community near the covered bridge included a building which was used for both school and a church.

In 1847, John Calvin Rece, who had been licensed to preach, began an effort for a new meetinghouse. It was erected in 1849, with the masonry being done by Luther E. Shepherd and a Mr. Swanson of Gallipolis, Ohio. Edmund C. Rece and Henry M. Johnson did the woodwork. Charles A.

Rece, then only eleven years old, hand planed every window and door jamb in the building.

When the Civil War broke out in 1861, Union troops occupied the church as a garrison, to protect the covered bridge which was guarded twenty-four hours a day.

On April 5, 1863, a group of Confederates under command of Captain P. M. Carpenter attempted to take the bridge, but were repulsed by the Union troops under a Captain Dove.

This bridge was replaced by a modern structure in recent years.

This relic of the past once spanned the Middle Fork River in Barbour County at Audra. When Audra State Park was built, it was torn down and replaced by an iron and concrete structure. From the files of the *Barbour Democrat* of October 1909 I learned that Harrison Kelley was employed by Lemuel Chenoweth, the "Bridge Builder of West Virginia," for fourteen years and assisted with building the Audra Bridge, in 1854-55, as well as the North Fork Bridge over the Hughes River, or Cairo Bridge. While hunting near the Middle Fork Bridge, Abraham Reger and John Harrow killed a bear with their hunting knives while it was in the act of killing their dogs. The above photo was loaned to me by Mrs. Pearl Nitz Kerr, Walled Lake, Michigan, formerly of Volga, West Virginia.

The covered bridge at Dent, Barbour County, was built in 1893. On February 14, 1923, the west end abutment collapsed. It was replaced by new abutment on June 23, 1923. In 1940 the old relic of horse and buggy days was torn down and a modern concrete structure built. I recall crossing this bridge many times as my sisters and I, accompanied by our neighbor girls, the Hardin sisters, rode horseback to attend spelling bees at the Garner Valley School or to attend an old-fashioned revival meeting at the Garner Valley Church. The horses would shy and prance as we approached the bridge but with a bit of urging they always proceeded across. The new bridge was built in 1940. (See next page.)

Dent Bridge, down.

This picturesque Hern's Mill Covered Bridge, over Milligan Creek, Greenbrier County, near the historic town of Lewisburg, was built in 1879. It is located on Route 40, about one mile from town. In 1962 it was extensively reconditioned and painted. It was in excellent condition in October of 1969 when I photographed it. It is one of the oldest covered bridges left in Greenbrier County. One other, at Hoke's Mill, has also survived new road and bridge construction.

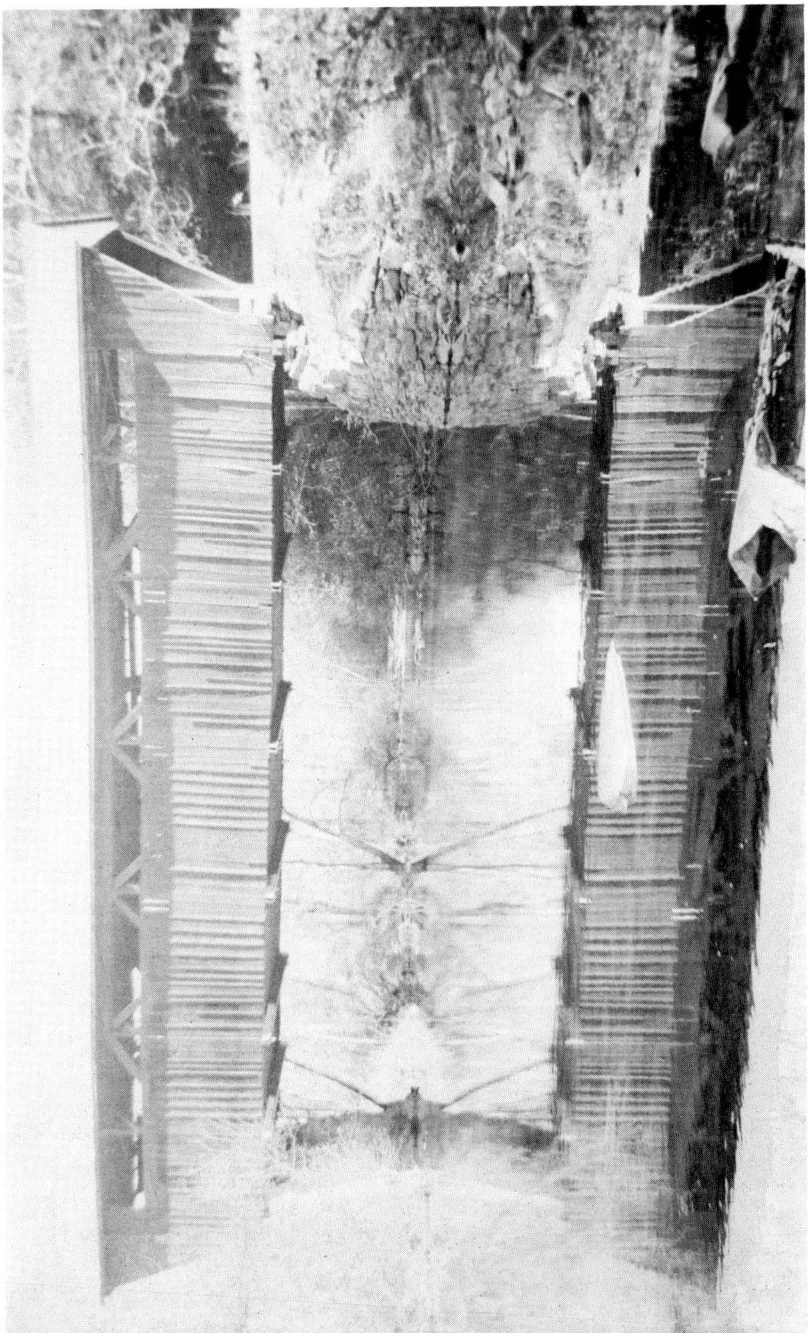

Hoke's Mill Covered Bridge.

HOKE'S MILL COVERED BRIDGE

Hoke's Mill Covered Bridge across Second Creek, in Greenbrier County, is between Neff's Orchards and Sinks Grove, on Route 62.

In 1957 when Greenbrier County completed a surfaced road from Lincoln to Hoke's Mill, this relic, nestling among large trees, with clear water murmuring over the rocks in Second Creek, was threatened. It was feared the old structure would be replaced; however, it is still standing and in good condition. To get this photograph I meant to go from stone to stone to the center of the creek. Best laid plans sometimes fail. One foot missed a rock, so I just waded on to the center. Being the latter part of October the temperature of the water had much to be desired. Fortunately I had extra shoes in my car, where a quick change was made.

Many old advertisements are still visible such as: "Murphy's All-American Shows" and "Use Foutz's Horse and Cattle Powders." There were also cards tacked up, showing that several Ohio historical societies had visited the bridge.

View of entrance to bridge can be seen on following page.

One of the most picturesque landmarks of horse and buggy days is this old covered bridge at Statts Mills, Jackson County, which spans the waters of the Tug Fork River. Built in 1882, it survived a great flood in 1913 and has served three generations of travelers. Passersby often see artists with easels set up sketching the still preserved relic. It is still in daily use. One of only two covered bridges left in Jackson County, it is located on Route 40 at junction of Secondary 34 North. It is one hundred feet in length, and still maintained by the State Department of Highways. See opposite page for another view of this bridge.

Covered bridge at Statts Mills.

Carrollton Covered Bridge.

CARROLLTON COVERED BRIDGE

About a mile from Route 119, on Route 36 at the junction of Secondary routes 11/3 East, nestles the much traveled Carrollton Covered Bridge. Built in the 1850s, it has served the communities of Mount Hebron, Audra and Carrollton, or Union as it is sometimes called. (See inside view of bridge on next page.)

Having served these communities in Barbour County for over a century, the bridge was in great need of repair. Citizens from the various communities made pleas by letters, telephone calls and visits to the district bridge headquarters of the State Road Commission at Weston, and the State Road Commission in Barbour County. Earth was piled at the entrances, and it was closed for a period of five months. Some of the siding was torn off and used to repair the Philippi bridge before its centennial celebration.

Happily, for all concerned, in 1962 the residents received the news that they would get a new $50,000 bridge and simultaneously get to keep their old one.

Sam G. McEwen, executive secretary and public relations officer for the State Road Commission, announced the decision on the new bridge. That decision was reached during a conference of Vincent J. Johnson, assistant state road commissioner; A. L. McMillion, maintenance director for the State Road Commission; and John E. Mayes, District Seven engineer at Weston.

The decision was to build a new bridge but to leave up the covering and sides of the old covered bridge, thus enabling the area residents to have their historic landmark and have the new bridge beneath the roof and between the sides of the old covered bridge.

Charles Wentz, of Boulder, recently informed me there was a covered bridge at Audra, built at about the same time as the Carrollton bridge. He and other workers sheltered in it, when working nearby, during the 1930s. It was torn down and replaced in the 1940s.

Early advertisements still visible.

The covered bridge over Paw Paw Creek, in Marion County, near Grant Town, is in fair condition. Forty-eight feet in length, it is located on routes 17/18 at the junction of Secondary 17 South. It is one of the older covered bridges of Marion County.

Covered bridge over Laurel Creek, Monroe County, twenty-five feet in length, and in good condition, is located on routes 23/4 at the junction of Secondary 219/11 West, near Lillydale.

The above covered bridge at Walkersville, in Lewis County, over the West Fork River, is in good condition. It is fifty-two and one-half feet in length. It can be seen from Route 44 at the junction of Secondary routes 44/2. I was unable to find the date it was built.

Booths Creek Bridge.

BOOTHS CREEK BRIDGE

Booths Creek is a stream of water, the left-hand fork of which heads in Harrison County, about two miles north of Bridgeport, and flows northward to Boothsville, from there northwesterly through Marion County into the West Fork. The first settler on the creek was James Booth who came there in 1771. The creek was named for him.

At one time there were two covered wooden bridges across the creek, one at Boothsville, now in Marion County, and across the line to Harrison County, and one across the line between Harrison and Taylor counties.

The lands along the creek in Harrison County are very fine and said to be "the garden spot" of the county. As the settlers increased, roads and bridges became a necessity.

On June 19, 1837, the County Court of Harrison County appointed Jesse Martin a commissioner to have the bridge then there repaired.

On May 20, 1838, Mr. Martin, Anderson Corbin and Absolom Knotts, who had been appointed commissioners to examine and report as to the expediency of building a bridge across the creek, reported, but upon consideration, the application for the bridge was denied by the court.

No other order about the bridge could be found, but the fact is well known that there was for many years a covered wooden bridge out of Boothsville, spanning the line into Harrison County.—Photo courtesy of Mrs. Josephine Little, Boothsville.

Covered bridge above Boothsville.

MORE ABOUT BOOTHS CREEK BRIDGES

This covered bridge was above Boothsville, about two miles, and across the line between Harrison and Taylor counties. The building of this bridge was a joint project by the two counties.

On June 14, 1881, the County Court of Harrison County appointed T. S. Spates a commissioner to meet with a commissioner appointed by the Taylor County Court, to take into consideration the building of a bridge across the creek on lands of E. L. Dixon and Gabe Smith. A report was to be made at the next term of court. However, no report was made until February 4, 1885, when Colonel Spates was appointed to examine a site and confer with a like commissioner from Taylor County, as to the desirability and probable cost, to be borne jointly by both counties, for building a bridge across the creek at the Gabe Smith farm.

On January 4, 1887, Spates was appointed commissioner for Harrison County and Elias Nixon by Taylor County for the construction of a covered wooden bridge at the joint expense of the two counties.

The commissioners reported that they had contracted with J. F. Sturm for building the bridge for $800, or $400 for each county.

The two covered bridges remained in use for many years but were finally replaced by iron structures.—Picture courtesy of Mrs. Josephine Little, Boothsville, taken in 1933. Mrs. Little and son are in the picture.

Barrackville Covered Bridge.

COVERED BRIDGE AT BARRACKVILLE COUNTY LANDMARK

One of the most familiar landmarks in Marion County is the old covered bridge at Barrackville which was constructed more than one hundred years ago.

According to records discovered in Richmond, Virginia, an agreement was signed April 10, 1853, between Austin Merrill and John and Robert McConnell for the stone work at $4.50 per perch. On April 25, 1853, an agreement was signed between Austin Merrill and Lemuel and Eli Chenoweth for construction of the superstructure at $12.50 per lineal foot.

The report of Merrill to the Board of Public Works of Virginia was made October 15, 1853, regarding the work, stating that it would be completed by November of that year.

The date of the actual completion of the bridge is not known. The dates mentioned were found by Milton C. Russell, who is in charge of the reference department of the General Library of Richmond, Virginia, for what is now West Virginia was part of the Commonwealth of Virginia when the bridge was built. A part of the information was discovered by Mrs. Lucy O'Neal of Philippi when she went to Richmond to seek information regarding the covered bridge at Philippi. Other information was obtained from a thesis by Calvin R. Conaway written in 1947.

The excellence of the work in the construction of this bridge is emphasized in the report of Austin Merrill to the Board of Public Works who stated:

"The material used in the construction of this work is of the very best quality and the work was done in a manner highly creditable to the contractors of both wood and stone. I think I hazard but little in saying this bridge, when completed, will have but few equals and perhaps not a superior in the Commonwealth of Virginia."

Merrill was also superintendent of construction of the Ohio River and Maryland Turnpike. Archie Graham, president of the Board of Public Works, in a report dated January 28, 1854, stated that "the Ohio River and Maryland Turnpike cannot be completed and consequently cannot accomplish the useful purposes of its projectors without additional appropriation."

Richard Sanders Allen, in writing of Lemuel Chenoweth, styles him a "master bridge builder." It seems fairly certain that Lemuel Chenoweth, with his flair for mathematics, fine woodworking and craftsmanship and perhaps a bit of "book knowledge" of Burr Truss bridges, designed his own covered wooden spans.

The Chenoweths were Virginia's and West Virginia's most famous bridge builders. The Barrackville bridge is 148 feet long and is a single span of the Burr-Arch type. Unlike the Philippi and Hunsaker bridges on the Fairmont and Beverly Turnpike, the Barrackville bridge was never a toll bridge.

Before 1852 the Chenoweths had built several bridges on the Staunton and Parkersburg Turnpike and the year before building the Barrackville bridge had built the Hunsaker bridge, 240 feet in length at Watson, and the Philippi bridge 280 feet long. All the technique and experience gained in building these bridges went far in making the Barrackville bridge the very finest. The West Virginia University School of Engineering has a model of the Barrackville bridge because it is considered one of the best examples of bridge construction ever observed for the period when it was built as well as for today.

Even at the time of the building of the bridge, Barrackville was an old and thriving town. Records show that several families had gone from here to Prickett's Fort at the time of the Revolution. Bailes Ice operated a gristmill and sawmill that drew trade from miles around. The Robisons had a hominy mill in operation not more than 100 yards from the north end of the bridge. John Barrack, the man from whom the town gets its name, and his sons, were busy shoeing horses and oxen, making ox yokes, and repairing wagons. Edmund Conaway's general store supplied to its customers dry goods, notions, salt, flour, some of which was hauled overland from Winchester, Virginia. The store was also the only voting place in the western part of the county except Fairmont. Prior to the building of the bridge, most of the people lived on the northwest side of Buffalo Creek in what is now known as the "old part" of Barrackville.

It seems almost a miracle that the bridge has survived war, fires, floods, and overloading for more than one hundred years. However, through its whole existence there have been certain citizens who always looked toward preserving it. Tradition tells us that Bailes and Dolly Ice prevailed on Gen. William E. Jones to spare the bridge at the time of "Jones's Raid" on April 29, 1863. For a long time after the building of the bridge, approximately twenty years, the timbers on the sides were exposed to the elements, not having been covered with weatherboarding at the time of original construction.

R. L. Cunningham, later a sheriff of Marion County, was given a contract to "side-up" the bridge. The late C. D. Conaway, while sheriff and later as county commissioner, looked after the preservation of the bridge by having it painted and a new roof put on. In 1934, C. A. Short of Shinnston was employed by the State Road Commission to repair the bridge. He built a sidewalk, repaired the approaches, reinforced the bridge truss and painted the bridge. The SRC again in 1951 did a lot to prolong the bridge's life by adding supports to the long stringers underneath.

Moatsville Covered Bridge before tragedy struck.

MOATSVILLE COVERED BRIDGE

The covered bridge at Moatsville, Barbour County, was built in 1893, according to Tommy Cline, ninety-three-year-old resident of Moatsville, with whom I talked April 27, 1972.

The contract for its construction was let to Jake Nicola, who owned a sawmill on the west side of Tygart Valley River.

All of the timbers were cut on his farm. The large stones for the abutments and center piers were also from along the bank of the river on his farm.

Some of the men who worked on the bridge were Charlie Wolfe, Dee Freeman, Alphus Freeman, and Criss Nicola.

When it was completed, it was dedicated by the residents of the community and surrounding areas with a square dance. He recalled they had two sets of dancers, swinging to the strains of the old fiddlers, at the same time.

From another friend, Mrs. E. S. Duckworth of Grafton, came the following information. "The old covered wooden bridge served for many different things than travel. The bridge was a favorite place for the youngsters of the community to gather at the east end near the railroad, as there were no shows or radios for their home entertainment. As many as 35 to 50 would gather in the evenings to play games such as 'Tag,' 'Snap,' and 'Strip-the-Willow.' Lanterns furnished the light. It was also a great place to go to see the G&B passenger train come and go. We always looked forward to Saturday evening in the summertime when we could meet friends and spend the evening in the old covered bridge."

My own recollection of the bridge was riding through it with my dad, driving cattle and sheep to the river farm, and driving through in the road wagon on the way to the farm to help with hay harvest.

In July 1912, tragedy struck the old bridge, when during a great flood on the Tygart Valley River the heavy currents carried her away. It was replaced by a modern concrete structure.—Photo courtesy Mrs. E. S. Duckworth, Grafton.

Dent's Run Covered Bridge.

The last remaining covered bridge in Monongalia County is over Dent's Run, north of Laurel Point. It is forty feet in length with five-ton load limit posted. Located on routes 43/4, it is in good condition and in use daily.

In 1962 Monongalia County had seven covered bridges standing.

See picture on facing page.

One of the well-known landmarks in Doddridge County was the covered bridge at West Union, over Middle Island Creek. It was designed by Lewis Wernwag and built during the construction of the Northwestern Turnpike. It served the county seat for over one hundred years. In 1941 it was "retired" and used for pedestrian traffic only. Its chances were good for lasting many years. Unhappily, in June 1950, during a sudden cloudburst, a fifteen-foot wall of water swept down the Middle Island Creek, destroying property and homes. The great wall of water washed the beautiful structure from its foundation and slammed it against the new concrete bridge that had succeeded it on U.S. Route 50. The State Road Commission had to dynamite the old hulk to relieve the pressure on the new span. The last of the West Union bridge floated in chunks down Middle Island Creek.—Photo by author.

Front view, Bulltown Covered Bridge.

BULLTOWN COVERED BRIDGE HISTORY

Captain Bull, the Delaware Indian chief whose name is perpetuated in Bulltown, on the Little Kanawha River in Braxton County, came to the hills of northwest Virginia as an exile from his homeland on the upper Susquehanna River in New York State. When he led his family and relatives to Bulltown, he was fleeing from the wrath of the English Indian commissioner. Whether Captain Bull and the Delawares planned to try to avenge their misfortune is not known, due to unforetold events that happened shortly after their arrival in the settlement.

They had planned to live here with the white man on terms of peace and friendship. The location was ideal. Game was plentiful and the rich fertile soil grew fine crops of corn. There were no other tribes to molest them. There were many salt springs near from which that much needed commodity could be made. Although no trace of the operations remains, Bulltown once had several salt furnaces, set up in 1810 by John Haymond. They supplied salt to a large part of northwest Virginia until the middle of the nineteenth century.

The murder of the Stroud family in Webster County by roving Shawnees, was seized upon by Jesse Hughes (who liked to kill Indians better than eat his dinner) as an excuse for an attack upon the innocent Delawares. The attack was made in the dead of night on the sleeping Indians. Hughes and his men killed and scalped every man, woman and child, and threw their bodies into the river.

A very substantial covered bridge was constructed across the Little Kanawha River at Bulltown, and notwithstanding its constant use for over half a century, carrying large bodies of troops, artillery, cavalry and thousands of heavily laden army wagons, the bridge was still in use in 1919. It was erected in 1850. It was still standing in 1941 but has since been torn down.

Pictures courtesy of Paul Currence, Napier, West Virginia, and Mrs. Ruth Skinner, Walkersville, West Virginia. See two pictures on following pages.

Bulltown Covered Bridge.

Neighbors visit just across the covered bridge at Bulltown in 1892.—Picture courtesy Mrs. Ruth Skinner, Walkersville.

The Salt Lick Bridge, Braxton County.—Picture courtesy Paul Currence, Napier, West Virginia.

Mrs. Ruth Skinner, Walkersville, West Virginia, loaned me the above picture and provided the following details about the bridge. The bridge was built in 1853 and 1854 by Harrison Kelley who was also employed in the building of several other bridges in the state. The Salt Lick bridge was a part of the Weston-Gauley Bridge Turnpike built during the early 1850s. The bridge was replaced by a steel structure at its original site in 1927. It was moved to Corley and reconstructed across Salt Lick at the Corley church. There it remained in use until 1939 when it was replaced with a modern structure. Note "Yankee jumper" and two sleighs in picture above.

This old covered bridge used to be seen on Route 7 at the junction of Secondary 22, at Berea, in Ritchie County, over the South Fork of the Hughes River. It was eighty-eight feet in length.

COVERED BRIDGE AT BEVERLY

As a part of the Staunton and Parkersburg Turnpike, a covered bridge was constructed across the Tygart Valley River at Beverly, Randolph County, in 1845. It was destroyed during the War Between the States. The bridge in the picture was built in 1873 to replace the earlier structure. Although records credit Lewis Wernwag with having designed the wooden-arch type covered bridge, it is generally believed in West Virginia that the second bridge was built by Lemuel Chenoweth of Beverly. Over a hundred years later, officials decided it was unsafe for school buses. Three charges of dynamite and a wrecking crew were necessary to demolish it. Had the officials known before what they knew after the blasting, the bridge might well have had a different fate. It was replaced by a modern structure around 1950.—Picture courtesy Mrs. Helen Harding, Beverly, West Virginia.

EPILOGUE

All told, there were hundreds of covered bridges erected in West Virginia, many while the region was still a part of the Old Dominion. Harrison County was one of the leading counties with the largest number. Fifty-seven of the familiar edifices graced its streams and rivers. Not all of them existed at one time, however. In 1888, a very disastrous flood removed most of the bridges of the West Fork River, including every bridge in the city of Clarksburg. Today, Harrison still reigns as covered bridge king, with three of the cherished landmarks remaining.

West Virginia's pioneer bridge builders, just as all other bridge builders, didn't cover their bridges to protect travelers, or to prevent horses from shying at the sight of water or to keep off snow. The main reason was that the roof and siding protected the supports and trusses from the elements. Floor planking was not so important, as it could be easily replaced. By protecting the wooden supports, the lives of the bridges were extended for many extra years.

All of us who remember the horse and buggy, remember far more pleasurable things about covered bridges. Here children played, the cool dust of the bridge floor tickling their bare feet, on hot summer days. Listening to the clip-clop of the horses' feet as they pranced through the bridge, whether we rode horseback or back of a spirited team pulling a buggy, surrey, or spring wagon.

It was a fishing platform as well as a shelter from sudden storms. In some cases it was where Sis's beau could steal a kiss on the way home from a party or church service. Many of them served as social gathering places, as I reported in the Moatsville Covered Bridge history.

Here, too, the traveling circus could post colorful posters informing folks far and near that the circus was coming to town, tobacco signs, and those advertising all kinds of "cure-alls." Some of these are still visible in the Carrollton and Milton bridges.

The covered bridges had their dark sides also, when a few travelers had to forfeit their purses to robbers. Granville

Davisson Hall, in his novel, *Daughter of the Elm*, describes a community in Harrison County, in which there was a well-organized band of thieves. In one chapter of the book, the author described a holdup in the famous Maulsby bridge, which fortunately did not result in any loot for the thieves. Their intended victim had deposited his money in a Clarksburg bank.

While wars, floods, fires, and just plain old age have taken their toll of Mountain State covered bridges, eighteen of these old relics are still standing. Most of them are in every-day use. The Philippi bridge still serves busy federal highway traffic twenty-four hours a day. Quite a record for a structure built 120 years ago.

The last remaining bridge designed by Lewis Wernwag, was the beautiful Cheat River bridge, Preston County. It had been preserved in its picturesque mountain setting as a state relic, to the delight of thousands of sightseers and picnickers. However, in 1964 fire destroyed it in twenty minutes. Only the large stone foundations remain.

Three of those still standing have undergone extensive repairs by the Department of Highways. Everything possible should be done to preserve these old relics of the past, loyal servants to the end.

Those remaining await your pen, ink, brush, or camera. Autumn is perhaps the best time of year to see them at their best. Surrounded by a rainbow of colors, they stand as they have for generations past, and hopefully, for generations to come. West Virginia's covered bridges—our familiar friends from the past.

ABOUT THE AUTHOR

Graduated West Liberty State College, 1918, Short Normal; 1926, Standard Normal; Fairmont State, 1946, A.B. degree. Postgraduate work, WVU. Taught first school at Locust Grove, one room school, where she started to school in 1905.

Taught in Barbour County until 1928, when she began in (and retired from) Taylor County. Had a 4-H Club in each school she taught in Barbour. Coming to Taylor schools she discovered they had no 4-H clubs. Feeling a need for such an organization she organized a club at Millerville. Soon other teachers desired clubs in their schools and asked her to help them get started. Worked with teachers in seven different communities, after school or at night meetings. This continued for seven years when the county court voted to hire a county agent. Was 4-H leader forty-five years. Attended first Volunteer Leaders Camp held at Jackson's Mill. Is life member of the 4-H All-Stars, state 4-H honorary organization for recognition of service to 4-H work.

Retired 1961. Hobbies, mostly woodwork.

PUBLISHER'S NOTE

To the reader of this book: If you have enjoyed this book tell your friends. With your help and that of others who read the book, an expanded second printing may be possible. If you have pictures of other bridges, not included here, please send them to Miss Auvil, who will have them copied for a possible second printing and then return pictures to you. If you can furnish her with histories of bridges no longer standing but do not have pictures, send this material to her, please. In some cases the pictures in this book were copied from old prints and are not good. Miss Auvil would appreciate it if persons with better pictures of these bridges would allow her to have them copied for possible future use. Mail information to Myrtle Auvil, Box 5, Grafton, West Virginia 26354. Thank you.